Lecture Notes in Physics

Edited by H. Araki, Kyoto, J. Ehlers, München, K. Hepp, Zürich
R. Kippenhahn, München, H. A. Weidenmüller, Heidelberg
and J. Zittartz, Köln
Managing Editor: W. Beiglböck

1–222

An Index
and
Other Useful Information

Springer-Verlag Berlin Heidelberg GmbH

ISBN 978-3-662-27889-5   ISBN 978-3-662-29391-1 (eBook)
DOI 10.1007/978-3-662-29391-1

© Springer-Verlag Berlin Heidelberg 1985
Originally published by Springer-Verlag Berlin Heidelberg New York Tokyo in 1985
31088/2.80.200. Be.
Offsetdruck: Julius Beltz, Hemsbach/Bergstr.

TABLE OF CONTENTS

I. Lecture Notes in Physics 1969 - 1985

In 1979 when the series celebrated its tenth anniversary I wrote:

Ten years ago the series **LECTURE NOTES IN PHYSICS** was founded with two goals in mind: Firstly, very good lectures, especially those devoted to themes of actual research, should become accessible to a wider audience, and secondly, proceedings of conferences should appear quickly and at a lower price. In both cases, however, preference was given to genuine "Lecture Notes", i.e., contrary to concise articles in scientific journals written for the expert, the contributions to this series should present many different aspects of the problem in question and should be intelligible for a wider audience quite often including students or workers from other fields of research. At the same time "Lecture Notes" also means that the form of a text might be tentative and less accurate than usually required for a textbook or a monograph. However, although tentative in form, a high scientific standard in correctness and in selection of the material presented is asked for.

Thanks to the generous help of Springer-Verlag, to the efforts of organizers of conferences, and last, but not least, to the splendid cooperation of the authors, these goals have been reached in a satisfactory way. Very rarely does a research paper in physics outlive a period of about three years, and therefore it is very gratifying to observe that some of the volumes of our series are still highly esteemed after a much longer lapse of time. As for the second goal, quite a few of the proceedings in this series appeared within weeks after the meeting in question. Due to a steadily increasing number of good manuscripts submitted it should be possible for the benefit of their readers to raise the standards further and to open this series to more areas of research in physics and its applications.

In only five years the number of volumes published has doubled from 111 in 1979 to 222 in February 1985. In an editorial meeting we analyzed this phenomenon and decided on a slight change in our policy. Contributions from astrophysics and new areas like scientific computing have recently shown a significant increase; a broadening of the scope of our series is highly welcome and will be pursued further. However, we also observe a less welcome increase in the ratio of proceedings to genuine lecture notes. Here a correction seems in order, and in the future the latter will be encouraged. Proceedings are still welcome but we will give priority to topical meetings of a review character rather than to collections of technical research papers of interest to the specialist only.

Again, I would like to take this opportunity to thank the editors of the series and the many referees for their help. Their usually very elaborate criticism also helped many of the authors to find a better balanced and a more satisfactory presentation of their ideas.

W. Beiglböck
Managing Editor

1. Series List

Customers from North America may place their order directly at Springer-Verlag New York, 175 Fifth Avenue, New York, NY 10010. In this case, please substitute the first four numbers of the ISBN (3-540) by 0-387.

Lecture Notes in Physics. Editors: Araki, H.; Ehlers, J.; Hepp, K.; Kippenhahn, J.; Weidenmüller, H. A.; Zittartz, J

Vol. 1: **Erdmann, J. C.:** Wärmeleitung in Kristallen. Theoretische Grundlagen und fortgeschrittene experimentelle Methoden. 1969. 76 Abb. II,283 Seiten. 535g ⟨3-540-04639-9⟩

Vol. 2: **Hepp, K.:** Théorie de la renormalisation. Cours donné à l'Ecole Polytechnique, Paris. 1969. III,215 pages. 425g ⟨3-540-04640-2⟩

Vol. 3: **Martin, A.:** Scattering Theory: Unitarity, Analyticity and Crossing. Notes taken by Schrader, R. 1969. IV,125 pages. Out of print ⟨3-540-04641-0⟩

Vol. 4: **Ludwig, G.:** Deutung des Begriffs physikalische Theorie und axiomatische Grundlegung der Hilbertraumstruktur der Quantenmechanik durch Hauptsätze des Messens. 1970. XI,469 Seiten. Vergriffen ⟨3-540-04941-X⟩

Vol. 5: **Schaaf, M.:** The Reduction of the Product of Two Irreducible Unitary Representations of the Proper Orthochronous Quantummechanical Poincaré Group. 1970. IV,120 pages. 260g ⟨3-540-05194-5⟩

Vol. 6: **Group Representations in Mathematics and Physics.** Battelle Seattle 1969 Rencontres. Editor: Bargmann, V. 1970. V,340 pages. 590g ⟨3-540-05310-7⟩

Vol. 7: **Lectures in Statistical Physics.** From the Advanced School for Statistical Mechanics and Thermodynamics. Austin, Texas, USA. By Balescu, R.; Lebowitz, J. L.; Prigogine, I.; Résibois, P.; Salsburg, Z. W. Compiled by Schieve, W. C.; Velarde, M. G.; Grecos, A. P. 1971. V,181 pages. Out of print ⟨3-540-05418-9⟩

Vol. 8: **Proceedings of the Second International Conference on Numerical Methods in Fluid Dynamics September 15-19, 1970, University of California, Berkeley.** Editor: Holt, M. 1971. IX,462 pages. Out of print ⟨3-540-05407-3⟩

Vol. 9: **Robinson, D. W.:** The Thermodynamic Pressure in Quantum Statistical Mechanics. 1971. Out of print ⟨3-540-05640-8⟩

Vol. 10: **Stewart, J. M.:** Non-Equilibrium Relativistic Kinetic Theory. 1971. III,113 pages. Out of print ⟨3-540-05652-1⟩

Vol. 11: **Steinmann, O.:** Perturbation Expansion in Axiomatic Field Theory. 1971. III,126 pages. Out of print ⟨3-540-05698-X⟩

Vol. 12: **Statistical Models and Turbulence.** Proceedings of a Symposium held at the University of California, San Diego (La Jolla) July 15-21, 1971. Editors: Rosenblatt, M.; Atta, C. van. (Reprint of the 1st edition) 1975. 129 figs. VIII,492 pages. 845g ⟨3-540-05716-1⟩

Vol. 13: **Ryan, M.:** Hamiltonian Cosmology. 1972. 16 figs. VII,169 pages. Out of print ⟨3-540-05741-2⟩

Vol. 14: **Methods of Local and Global Differential Geometry in General Relativity.** Proceedings of the Regional Conference on Relativity held at the University of Pittsburgh, Pittsburgh, Pennsylvania, July 13-17, 1970. Editors: Farnsworth, D.; Fink, J.; Porter, J.; Thompson, A. 1972. 33 figs. V,188 pages. Out of print ⟨3-540-05793-5⟩

Vol. 15: **Fierz, M.:** Vorlesungen zur Entwicklungsgeschichte der Mechanik. 1972. V,97 Seiten. 200g ⟨3-540-05907-5⟩

Vol. 16: **Georgii, H. O.:** Phasenübergang 1. Art bei Gittergasmodellen. Klassische Systeme gleichartiger Teilchen mit paarweiser Wechselwirkung. 1972. IX,167 Seiten. 315g ⟨3-540-06025-1⟩

Vol. 17: **Strong Interaction Physics.** International Summer Institute on Theoretical Physics in Kaiserslautern 1972. Editors: Rühl, W.; Vancura, A. 1973. Out of print ⟨3-540-06141-X⟩

Vol. 18: **Proceedings of the Third International Conference on Numerical Methods in Fluid Mechanics.** Vol. 1: General Lectures, Fundamental Numerical Techniques. July 3 - 7, 1972, Universities of Paris VI and XI. Editors: Cabannes, H.; Temam, R. 1973. Out of print MA ⟨3-540-06170-3⟩

Vol. 19: **Proceedings of the Third International Conference on Numerical Methods in Fluid Mechanics.** Vol. 2: Problems of Fluid Mechanics. July 3-7, 1972, Universities of Paris VI and XI. Editors: Cabannes, H.; Temam, R. 1973. 175 figs. VII,275 pages. Out of print
MA ⟨3-540-06171-1⟩

Vol. 20: **Statistical Mechanics and Mathematical Problems.** Battelle Seattle 1971 Rencontres. Editor: Lenard, A. 1973. 3 figs. VIII,247 pages. Out of print ⟨3-540-06194-0⟩

Vol. 21: **Optimization and Stability Problems in Continuum Mechanics.** Lectures Presented at the Symposium on Optimization and Stability Problems in Continuum Mechanics, Los Angeles, California, August 24, 1971. Editor: Wang, P. K. C. 1973. 14 figs. V,94 pages. 195g
⟨3-540-06214-9⟩

Vol. 22: **Proceedings of the Europhysics Study Conference on Intermediate Processes in Nuclear Reactions.** August 31 - September 5, 1972 Plitvice Lakes, Yugoslavia. Editors: Cindro, N.; Kulisic, P.; Mayer-Kuckuk, T. 1973. 142 figs. XIV,329 pages. 590g ⟨3-540-06526-1⟩

Vol. 23: **Nuclear Structure Physics.** Proceedings of the Minerva Symposium on Physics held at the Weizmann Institute of Science, Rehovot, Israel, April 2-5, 1973. Editors: Smilansky, U.; Talmi, I.; Weidenmüller, H. A. 1973. 174 figs. XII,296 pages. 545g
⟨3-540-06554-7⟩

Vol. 24: **Snipes, R. F.:** Statistical Mechanical Theory of the Electrolytic Transport of Non-electrolytes. 1973. 3 figs. V,210 pages. 380g
⟨3-540-06566-0⟩

Vol. 25: **Constructive Quantum Field Theory.** The 1973 "Ettore Majorana" International School of Mathematical Physics. Editors: Velo, G.; Wightman, A. S. 1973. 575g
⟨3-540-06608-X⟩

Vol. 26: **Hubert, A.:** Theorie der Domänenwände in geordneten Medien. 1974. XII,377 Seiten. 660g ⟨3-540-06680-2⟩

Vol. 27: **Zeytounian, R. K.:** Notes sur les Ecoulements Rotationnels de Fluides Parfaits. 1974. XIII,407 pages. 610g ⟨3-540-06721-3⟩

Vol. 28: **Lectures in Statistical Physics.** Advanced School for Statistical Mechanics and Thermodynamics, Austin, Texas, USA. By Ehlers, J.; Ford, J.; George, C.; Miller, R.; Montroll, E.; Schieve, W. C.; Turner, J. S. Editors: Schieve, W. C.; Turner, J. S. 1974. 76 figs. VI,342 pages. 595g ⟨3-540-06711-6⟩

Vol. 29: **Foundations of Quantum Mechanics and Ordered Linear Spaces.** Advanced Study Institute, Marburg 1973. Editors: Hartkämper, A.; Neumann, H. 1974. VI,355 pages. Out of print ⟨3-540-06725-6⟩

Vol. 30: **Polarization Nuclear Physics.** Proceedings of a Meeting held at Ebermannstadt, Germany, October 1-5, 1973. Editor: Fick, D. 1974. 130 figs. IX,292 pages. 500g
⟨3-540-06978-X⟩

Vol. 31: **Transport Phenomena.** Sitges International School of Statistical Mechanics, Barcelona, June 1974. Editors: Kirczenow, G.; Marro, J. 1974. 51 figs. XIV,517 pages. 895g
⟨3-540-06955-0⟩

Vol. 32: **Particles, Quantum Fields and Statistical Mechanics.** Proceedings of the 1973 Summer Institute in Theoretical Physics held at Mexico City. Editors: Alexanian, M.; Zepeda, A. 1975. 16 figs. V,132 pages. 245g
⟨3-540-07022-2⟩

Vol. 33: **Classical and Quantum Mechanical Aspects of Heavy Ion Collisions.** Symposium held at the Max-Planck-Institut für Kernphysik Heidelberg, Germany, October 2-5, 1974. Editors: Harney, H. L.; Braun-Munzinger, P.; Gelbke, C. K. 1975. 206 figs. VII,311 pages. 555g ⟨3-540-07025-7⟩

Vol. 34: **One Dimensional Conductors.** GPS Summer School Proceedings. Editor: Schuster, H. G. 1975. 105 figs. X,371 pages. Out of print ⟨3-540-07024-9⟩

Vol. 35: **Proceedings of the Fourth International Conference on Numerical Methods in Fluid Dynamics.** University of Colorado, June 24-28, 1974. Editor: Richtmyer, R. D. 1975. Numerous figs. V,457 pages. 790g
⟨3-540-07139-3⟩

Vol. 36: **Gatignol, R.:** Théorie Cinétique des Gaz à Répartition Discrète de Vitesses. 1975. II,219 pages. 390g ⟨3-540-07156-3⟩

Vol. 37: **Trends in Elementary Particle Theory.** International Summer Institute on Theoretical Physics in Bonn 1974. Editors: Rollnik, H.; Dietz, K. 1975. V,472 pages. Out of print
⟨3-540-07160-1⟩

Vol. 38: **Dynamical Systems, Theory and Applications.** Battelle Seattle 1974 Rencontres. Editor: Moser, J. 1975. VI,624 pages. Out of print ⟨3-540-07171-7⟩

Vol. 39: **International Symposium on Mathematical Problems in Theoretical Physics.** January 23-29, 1975, Kyoto University, Kyoto, Japan. Editor: Araki, H. 1975. XII,562 pages. 960g ⟨3-540-07174-1⟩

Vol. 40: **Effective Interactions and Operators in Nuclei.** Proceedings of the Tucson International Topical Conference on Nuclear Physics, held at the University of Arizona, Tucson, June 2-6, 1975. Editor: Barrett, B. R. 1975. XII,339 pages. 595g ⟨3-540-07400-7⟩

Vol. 41: **Progress in Numerical Fluid Dynamics.** Lecture Series held at the von Karman Institute for Fluid Dynamics, Rhode-St.-Genese, Belgium, February 11-15, 1974. Revised and Updated Version. Editor: Wirz, H. J. 1975. 146 figs. 6 tab. V,471 pages. 805g
⟨3-540-07408-2⟩

Vol. 42: **H II Regions and Related Topics.** Proceedings of a Symposium held at Mittelberg, Kleinwalsertal, Austria, January 13-17, 1975. Editors: Wilson, T. L.; Downes, D. 1975. 140 figs., 31 tables, XII,488 pages. 845g
⟨3-540-07409-0⟩

Vol. 43: **Laser Spectroscopy.** Proceedings of the 2nd International Conference, Megève, France, June 23-27, 1975. Editors: Haroche, S.; Pebay-Peyroula, J. C.; Hänsch, T. W.; Harris, E. E. 1975. 230 figs., 30 tables, X,468 pages (5 pages in French). Out of print
⟨3-540-07411-2⟩

Vol. 44: **Breuer, R. A.:** Gravitational Perturbation Theory and Synchrotron Radiation. 1975. 20 figs., 6 tables. VI,196 pages. 340g
⟨3-540-07530-5⟩

Vol. 45: **Dynamical Concepts on Scaling Violation and the New Resonances in $e^+ e^-$ Annihilation.** Editor: Humpert, B. 1976. VII,248 pages. 455g ⟨3-540-07539-9⟩

Vol. 46: **Flaherty, E. J.:** Hermitian and Kählerian Geometry in Relativity. 1976. 16 figs., 2 tables. VIII,365 pages. Out of print
⟨3-540-07540-2⟩

Vol. 47: **Padé Approximants Method and Its Applications to Mechanics.** Editor: Cabannes, H. 1976. 1 portrait, 54 figs. 18 tab. XV,267 pages. Out of print ⟨3-540-07614-X⟩

Vol. 48: **Interplanetary Dust and Zodiacal Light.** Proceedings of the IAU-Colloquium No. 31, Heidelberg, June 10-13, 1975. Editors: Elsässer, H.; Fechtig, H. 1976. 165 figs. 44 tab. XII,496 pages. 855g ⟨3-540-07615-8⟩

Vol. 49: **Harter, W. G.; Patterson, C. W.:** A Unitary Calculus for Electronic Orbitals. 1976. 15 figs. 6 tab. XII,144 pages. 285g
⟨3-540-07699-9⟩

Vol. 50: **Group Theoretical Methods in Physics.** Fourth International Colloquium, Nijmegen, NL. 1975. Editors: Janner, A.; Janssen, T.; Boon, M. 1976. 16 figs. 18 tab. XIII,629 pages (47 pages in French). Out of print
⟨3-540-07789-8⟩

Vol. 51: **Nörenberg, W.; Weidenmüller, H.-A.:** Introduction to the Theory of Heavy-Ion Collisions. 3rd ed. planned as monograph

Vol. 52: **Mladjenović, M.:** Development of Magnetic β-Ray Spectroscopy. 1976. X,282 pages. Out of print ⟨3-540-07851-7⟩

Vol. 53: **Simms, D. J.; Woodhouse, N. M. J.:** Lectures on Geometric Quantization. 1976. V,166 pages. 310g ⟨3-540-07860-6⟩

Vol. 54: **Critical Phenomena.** Sitges International School on Statistical Mechanics, June 1976, Sitges, Barcelona/Spain. Editors: Brey, J.; Jones, R. B. Director: Garrido, L. 1976. 50 figs. XI,383 pages. Out of print
⟨3-540-07862-2⟩

Vol. 55: **Nuclear Optical Model Potential.** Proceedings of the Meeting held in Pavia, April 8-9, 1976. Editors: Boffi, S.; Passatore, G. 1976. 62 figs. 6 tab. VI,221 pages. 400g
⟨3-540-07864-9⟩

Vol. 56: **Current Induced Reactions.** International Summer Institute on Theoretical Particle Physics in Hamburg 1975. Editors: Körner, J. G.; Kramer, G.; Schildknecht, D. 1976. 147 figs. 13 tab. V,553 pages. 440g
⟨3-540-07866-5⟩

Vol. 57: **Physics of Highly Excited States in Solids.** Proceedings of the 1975 Oji Seminar at Tomakomai, Japan, September 9-13, 1975. Editors: Ueta, M.; Nishina, Y. 1976. 138 figs. 3 tab. IX,391 pages. 680g ⟨3-540-07991-2⟩

Vol. 58: **Computing Methods in Applied Sciences.** Second International Symposium, December 15-19, 1975. Editors: Glowinski, R.; Lions, J. L. 1976. 188 figs., 15 tab. VIII,593 pages (191 pages in French) 1010g
⟨3-540-08003-1⟩

Vol. 59: **Proceedings of the Fifth International Conference on Numerical Methods in Fluid Dynamics.** June 28 - July 3, 1976, Twente University, Enschede. Editors: Vooren, A. I. van de; Zandbergen, P. J. 1976. 280 figs., 18 tables. VII,459 pages. 790g   ⟨3-540-08004-X⟩

Vol. 60: **Gruber, C.; Hintermann, A.; Merlini, D.:** Group Analysis of Classical Lattice Systems. 1977. 49 figs., 5 tables. XIV,326 pages. 785g   ⟨3-540-08137-2⟩

Vol. 61: **Photonuclear Reactions I.** International School on Electro- and Photonuclear Reactions, Erice Italy 1976. Editors: Costa, S.; Schaerf, C. 1977. 362 figs., 62 tab. VII,650 pages. 1105g   ⟨3-540-08139-9⟩

Vol. 62: **Photonuclear Reactions II.** International School on Electro- and Photonuclear Reactions, Erice, Italy 1976. Editors: Costa, S.; Schaerf, C. 1977. 174 figs., 13 tab. VII,301 pages. 505g   ⟨3-540-08140-2⟩

Vol. 63: **Harmonic Analysis** on the n-Dimensional Lorentz Group and Its Application to Conformal Quantum Field Theory. By Dobrev, V. K.; Mack, G.; Petkova, V. B.; Petrova, S. G.; Todorov, I. T. 1977. 6 figs. X,280 pages. Out of print   ⟨3-540-08150-X⟩

Vol. 64: **Waves on Water of Variable Depth.** Proceedings of a symposium held under the auspices of the International Union of Theoretical and Applied Mechanics (IUTAM) and the Australian Academy of Science at the Academy in Canberra, 1976. Editors: Provis, D. G.; Radok, R. 1977. 134 figs., 10 tab. 231 pages. 410g   AA ⟨3-540-08253-0⟩

Vol. 65: **Organic Conductors and Semiconductors.** Proceedings of the International Conference, Siófok, Hungary, 1976. Editors: Pál, L.; Grüner, G.; Jánossy, A.; Sólyom, J. 1977. 167 figs., 14 tab. 654 pages. 1165g
AO ⟨3-540-08255-7⟩

Vol. 66: **Völkel, A. H.:** Fields, Particles, and Currents. 1977. VI,354 pages. 620g
⟨3-540-08347-2⟩

Vol. 67: **Drechsler, W.; Mayer, M. E.:** Fibre Bundle Techniques in Gauge Theories. Lectures in Mathematical Physics at the University of Texas at Austin. Edited by Böhm, A.; Dollard, J. D. 1977. IX,248 pages. Out of print
⟨3-540-08350-2⟩

Vol. 68: **Venkatesh, Y. V.:** Energy Methods in Time-Varying System Stability and Instability Analyses. 1977. 2 figs. IX,256 pages. 465g
⟨3-540-08430-4⟩

Vol. 69: **Rohlfs, K.:** Lectures on Density Wave Theory. 1977. 76 figs., 14 tab. VI,184 pages. Out of print   ⟨3-540-08448-7⟩

Vol. 70: **Wave Propagation and Underwater Acoustics.** Editors: Keller, J. B.; Papadakis, J. S. 1977. 32 figs. VIII,287 pages. 510g
⟨3-540-08527-0⟩

Vol. 71: **Problems of Stellar Convection.** Proceedings of the Colloquium No. 38 of the International Astronomical Union held in Nice, 16-20 August, 1976. Editors: Spiegel, E. A.; Zahn, J. P. 1977. 76 figs., 1 tab. VIII,363 pages. 635g   ⟨3-540-08532-7⟩

Vol. 72: **Les instabilités hydrodynamique en convection Libre, forcée et mixte.** Edité par Legros, J.-C.; Platten, J. K. 1978. 71 ills., 15 tabs. IX,202 pages (52 pages en Anglais) 375g
⟨3-540-08652-8⟩

Vol. 73: **Invariant Wave Equations.** Proceedings of the "Ettore Majorana" International School of Mathematical Physics, held in Erice, Italy, June 27 to July 9, 1977. Editors: Velo, G.; Wightman, A. S. 1978. V,416 pages. 715g
⟨3-540-08655-2⟩

Vol. 74: **Collet, P.; Eckmann, J. P.:** A Renormalization Group Analysis of the Hierarchical Model in Statistical Mechanics. 1978. 11 figs. III,199 pages. Out of print   ⟨3-540-08670-6⟩

Vol. 75: **Structure and Mechanisms of Turbulence I.** Proceedings of the Symposium on Turbulence held at the Technische Universität Berlin, August 1-5, 1977. Editor: Fiedler, H. 1978. 210 figs., 5 tab. XX,295 pages. 550g
⟨3-540-08765-6⟩

Vol. 76: **Structure and Mechanisms of Turbulence II.** Proceedings of the Symposium on Turbulence held at the Technische Universität Berlin, August 1-5, 1977. Editor: Fiedler, H. 1978. 209 figs., 3 tab. XX,406 pages. 725g
⟨3-540-08767-2⟩

Vol. 77: **Topics in Quantum Field Theory and Gauge Theories.** Proceedings of the VIII International Seminar on Theoretical Physics, held by G.I.F.T., in Salamanca, June 13-19, 1977. Editor: Azcárraga, J. A. de. 1978. 60 figs., 14 tab. X,378 pages. 660g ⟨3-540-08841-5⟩

Vol. 78: **Böhm, A.:** The Rigged Hilbert Space and Quantum Mechanics. Lectures in Mathematical Physics at the University of Texas at Austin. Editors: Böhm, A.; Dollard, J. D. 1978. IX,70 pages. Out of print ⟨3-540-08843-1⟩

Vol. 79: **Group Theoretical Methods in Physics.** Sixth International Colloquium Tübingen 1977. Editors: Kramer, P.; Rieckers, A. 1978. 24 figs., 18 tab. XVIII,546 pages. 945g ⟨3-540-08848-2⟩

Vol. 80: **Mathematical Problems in Theoretical Physics.** International Conference held in Rome, June 6-15, 1977. Editors: Dell-Antonio, G.; Doplicher, S.; Jona-Lasinio, R. 1978. 9 figs. 1 tab. VI,438 pages. 750g ⟨3-540-08853-9⟩

Vol. 81: **MacGregor, M. H.:** The Nature of the Elementary Particle. 1978. 101 figs., 39 tab. XXII,482 pages. 855g ⟨3-540-08857-1⟩

Vol. 82: **Few Body Systems and Nuclear Forces I.** 8. International Conference held in Graz, August 24-30, 1978. Editors: Zingl, H.; Haftel, M.; Zankel, H. 1978. 235 figs., 43 tab. XIX,442 pages. 780g ⟨3-540-08917-9⟩

Vol. 83: **Experimental Methods in Heavy Ion Physics.** Editor: Bethge, K. 1978. 89 figs., 27 tab. V,251 pages. 445g ⟨3-540-08931-4⟩

Vol. 84: **Stochastic Processes in Nonequilibrium Systems.** Sitges International School of Statistical Mechanics, June 1978; Sitges, Barcelona/Spain. Editors: Garrido, L.; Seglar, P.; Shepherd, P. J. 1978. 35 figs., 5 tab. XI,352 pages. 630g ⟨3-540-08942-X⟩

Vol. 85: **Applied Inverse Problems.** Lectures presented at the RCP 264 "Etude Interdisciplinaire des Problèmes Inverses", sponsored by the Centre National de la Recherche Scientifique. Editor: Sabatier, P. C. 1978. 37 figs., 13 tab. V,425 pages. (89 in French) 730g ⟨3-540-09094-0⟩

Vol. 86: **Few Body Systems and Electromagnetic Interaction.** Proceedings of the workshop held in Frascati (Italy), March 7-10, 1978. Editors: Ciofi Degli Atti, C.; De Sanctis, E.. With contributions by numerous experts. 1978. 172 figs., 26 tab. VI,352 pages. 610g ⟨3-540-09095-9⟩

Vol. 87: **Few Body Systems and Nuclear Forces II.** 8. International Conference held in Graz, August 24-30, 1978. Editors: Zingl, H.; Haftel, M.; Zankel, H. 1978. 346 figs., 30 tab. X,545 pages. 935g ⟨3-540-09099-1⟩

Vol. 88: **Hutter, K.; Ven, A. A. F. van de:** Field Matter Interactions in Thermoelastic Solids. A Unification of Existing Theories of Elektro-Magneto-Mechanical Interactions. 1978. VIII,231 pages. 420g ⟨3-540-09105-X⟩

Vol. 89: **Microscopic Optical Potentials.** Proceedings of the Hamburg Topical Workshop on Nuclear Physics, held at the University of Hamburg, Hamburg, Germany, September 25-27, 1978. Editor: Geramb, H. V. v. With 42 contributions by numerous experts. 1979. 270 figs., 30 tab. XI,481 pages. Out of print ⟨3-540-09106-8⟩

Vol. 90: **Sixth International Conference on Numerical Methods in Fluid Dynamics.** Proceedings Tbilisi (U.S.S.R.) June 21-24, 1978. Editors: Cabannes, H.; Holt, M.; Rusanov, V. V. 1979. 405 figs., 11 tab. VIII,620 pages. (9 pages in French) 1110g ⟨3-540-09115-7⟩

Vol. 91: **Computing Methods in Applied Sciences and Engineering, 1977, II.** Third International Symposium, Dec. 5-9, 1977, Iria Laboria Institut de Recherche d'Informatique et d'Automatique. Editors: Glowinski, R.; Lions, J. L. 1979. 113 figs., 4 tab. VI,359 pages (108 pages in French) (Part 1 see Lecture Notes in Mathematics, Vol. 704). Out of print ⟨3-540-09119-X⟩

Vol. 92: **Nuclear Interactions.** Conference held in Canberra, 28 August - 1 Sept. 1978 under the auspices of the International Union of Pure and Applied Physics, the Australian Institute of Physics and the Australian Academy of Science. Editor: Robson, B. A. 1979. 563 figs., 52 tab. XXIV,507 pages. 890g ⟨3-540-09102-5⟩

Vol. 93: **Stochastic Behavior in Classical and Quantum Hamiltonian Systems.** Volta Memorial Conference, Como 1977. Editors: Casati, G.; Ford, J. With contributions by numerous experts. 1979. 81 figs., 11 tab. VI,375 pages. 650g ⟨3-540-09120-3⟩

14

Vol. 94: **Group Theoretical Methods in Physics.** Seventh International Colloquium and Integrative Conference on Group Theory and Mathematical Physics, Austin, Texas, September 11-16, 1978. Editors: Beiglböck, W.; Böhm, A.; Takasugi, E. 1979. 26 figs., 3 tab. XIII,540 pages. 935g ⟨3-540-09238-2⟩

Vol. 95: **Quasi One-Dimensional Conductors I.** Proceedings of the International Conference Dubrovnik, SR Croatia, SFR Yugoslavia, 1978. Editors: Barišić, S.; Bjeliš, A.; Cooper, J. R.; Leontić, B. 1979. 192 figs., 25 tab. X,371 pages. 650g ⟨3-540-09240-4⟩

Vol. 96: **Quasi One-Dimensional Conductors II.** Proceedings of the International Conference Dubrovnik SR Croatia, SFR Yugoslavia, 1978. Editors: Barišić, S.; Bjeliš, A.; Cooper, J. R.; Leontič, B. 1979. 172 figs., 24 tab. XII,461 pages. 760g ⟨3-540-09241-2⟩

Vol. 97: **Hughston, L. P.:** Twistors and Particles. 1979. 1 fig., 9 tab. VIII,153 pages. 290g ⟨3-540-09244-7⟩

Vol. 98: **Nonlinear Problems in Theoretical Physics.** Proceedings of the IX G.I.F.T. International Seminar on Theoretical Physics, held at Jaca, Huesca (Spain), June 1978. Editor: Ranada, A. F. 1979. 16 figs., 1 tab. X,216 pages. 395g ⟨3-540-09246-3⟩

Vol. 99: **Drieschner, M.:** Voraussage - Wahrscheinlichkeit - Objekt. Über die begrifflichen Grundlagen der Quantenmechanik. 1979. 8 Abb., 3 Tab. XI,308 Seiten. 550g ⟨3-540-09248-X⟩

Vol. 100: **Einstein Symposion Berlin.** Aus Anlass der 100. Wiederkehr seines Geburtstages 25. bis 30. März 1979. Hrsg.: Nelkowski, H.; Hermann, A.; Poser, H.; Schrader, R.; Seiler, R. With contributions by numerous experts. 1979. 35 figs., 6 tab. VII,550 pages (273 pages in German) 935g ⟨3-540-09718-X⟩

Vol. 101: **Martin-Löf, A.:** Statistical Mechanics and the Foundations of Thermodynamics. 1979. V,120 pages. 230g ⟨3-540-09255-2⟩

Vol. 102: **Hora, H.:** Nonlinear Plasma Dynamics at Laser Irradiation. Notes from lectures presented during the winter semester 1978/79 at the Department of Laser Physics, Institute of Applied Physics, University Berne, Switzerland. Editorial assistance: Schwarzenbach, P. 1979. 89 figs. VIII,242 pages. 440g ⟨3-540-09502-0⟩

Vol. 103: **Martin, P. A.:** Modèles en Mécanique Statistique des Processus Irreversibles. Cours organisé par le Troisième Cycle de la Physique en Suisse Romande. 1979. 16 ills. IV,134 pages. 260g ⟨3-540-09509-8⟩

Vol. 104: **Dynamical Critical Phenomena and Related Topics.** Proceedings of the International Conference, held at the University of Geneva, Switzerland, April 2-6, 1979. Editor: Enz, C. P. 1979. 105 figs., 3 tab. XII,390 pages. 680g ⟨3-540-09523-3⟩

Vol. 105: **Dynamics and Instability of Fluid Interfaces.** Proceedings of a meeting, held at the Technical University of Denmark, Lyngby, May 1978. Editor: Sorensen, T. S. 1979. V,315 pages. 550g ⟨3-540-09524-1⟩

Vol. 106: **Feynman Path Integrals.** Proceedings of the International Colloquium, Held in Marseille, May 1978. Editors: Albeverio, S.; Combe, P.; Høegh-Krohn, R.; Rideau, G.; Sirugue-Collin, M.; Sirugue, M.; Stora, R. 1979. 20 figs., 3 tab. XI,451 pages (36 pages in French). Out of print ⟨3-540-09532-2⟩

Vol. 107: **Kijowski, J.; Tulczyjew, W. M.:** A Symplectic Framework for Field Theories. 1979. IV,257 pages. 455g ⟨3-540-09538-1⟩

Vol. 108: **Nuclear Physics with Electromagnetic Interactions.** Proceedings of the International Conference, Held in Mainz, Germany, June 5-9, 1979. Editors: Arenhövel, H.; Drechsel, D. 1979. 400 figs., 48 tab. IX,509 pages. Out of print ⟨3-540-09539-X⟩

Vol. 109: **Physics of the Expanding Universe.** Cracow School on Cosmology, Jodowy Dwor, September 1978 Poland. Editor: Demiánski, M. 1979. 39 figs., 15 tab. V,210 pages. 390g ⟨3-540-09562-4⟩

Vol. 110: **Park, D.:** Classical Dynamics and Its Quantum Analogues. 1979. 85 figs., 2 tab. VIII,339 pages. 595g ⟨3-540-09565-9⟩

Vol. 111: **Schmidt, H. J.:** Axiomatic Characterization of Physical Geometry. 1979. 27 figs. V,163 pages. 310g ⟨3-540-09719-8⟩

Vol. 112: **Imaging Processes and Coherence in Physics.** Proceedings of a workshop, held at the Centre de Physique, Les Houches, France, March 1979. Editors: Schlenker, M.; Fink, M.; Goedgebuer, J.-P.; Malgrange, C.; Vienot, J.-C.; Wade, R. H. 1980. 327 figs., 17 tab. XIX,577 pages. 1000g ⟨3-540-09727-9⟩

Vol. 113: **Recent Advances in the Quantum Theory of Polymers.** Proceedings of the workshop held in Namur (Belgium), February 11-14, 1979. Editors: André, J.-M.; Brédas, J.-L.; Delhalle, J.; Ladik, J.; Leroy, G.; Moser, C. 1980. 100 figs., 35 tab. V,306 pages. 545g
⟨3-540-09731-7⟩

Vol. 114: **Stellar Turbulence.** Proceedings of Colloquium 51 of the International Astronomical Union, held at the University of Western Ontario, London, Ontario, Canada, August 27-30, 1979. Editors: Gray, D. F.; Linsky, J. L. 1980. 97 figs., 8 tab. IX,308 pages. 545g
⟨3-540-09737-6⟩

Vol. 115: **Modern Trends in the Theory of Condensed Matter.** Proceedings of the XVI Karpacz Winter School of Theoretical Physics, February 19 - March 3, 1979, Karpacz, Poland. Editors: Pekalski, A.; Przystawa, J. 1980. 164 figs., 3 tab. IX,597 pages. 1020g
⟨3-540-09752-X⟩

Vol. 116: **Mathematical Problems in Theoretical Physics.** Proceedings of the International Conference on Mathematical Physics Lausanne, Switzerland, August 20-25, 1979. Editor: Osterwalder, K. 1980. 42 figs., 9 tab. VIII,412 pages. 715g
⟨3-540-09964-6⟩

Vol. 117: **Deep Inelastic and Fusion Reactions with Heavy Ions.** Proceedings of the Symposium at the Hahn-Meitner-Institut für Kernforschung, Berlin, October 23-25, 1979. Editor: Oertzen, W. von. 1980. 234 figs., 11 tab. XIII,394 pages. 700g
⟨3-540-09965-4⟩

Vol. 118: **Quantum Chromodynamics.** Proceedings of the X G.I.F.T. International Seminar on Theoretical Physics, held at Jaca, Huesca (Spain) June 1979. Editors: Alonso, J. L.; Tarrach, R. 1980. 150 figs. IX,424 pages. 735g
⟨3-540-09969-7⟩

Vol. 119: **Nuclear Spectroscopy.** Lecture Notes of the workshop held at Gull Lake, Michigan, August 27 - September 7, 1979. Editors: Bertsch, G. F.-; Kurath, D. 1980. 79 figs., 4 tab. VII,250 pages. 440g ⟨3-540-09970-0⟩

Vol. 120: **Nonlinear Evolution Equations and Dynamical Systems.** Proceedings of the meeting held at the University of Lecce, June 20-23, 1979. Editors: Boiti, M.; Pempinelli, F.; Soliani, G. 1980. 25 figs. VI,368 pages. 635g
⟨3-540-09971-9⟩

Vol. 121: **Wiegel, F. W.:** Fluid Flow Through Porous Macromolecular Systems. 1980. 4 figs., 6 tab. V,102 pages. 205g ⟨3-540-09973-5⟩

Vol. 122: **New Developments in Semiconductor Physics.** Proceedings of the International Summer School, Held in Szeged, Hungary, July 1-6, 1979. Editors: Beleznay, F.; Ferenczi, G.; Giber, J. 1980. 171 figs., 23 tab. V,276 pages. Out of print ⟨3-540-09988-3⟩

Vol. 123: **Mayer, D. H.:** The Ruelle-Araki Transfer Operator in Classical Statistical Mechanics. 1980. VIII,154 pages. 295g
⟨3-540-09990-5⟩

Vol. 124: **Gravitational Radiation, Collapsed Objects, and Exact Solutions.** Proceedings of the Einstein Centenary Summer School, held in Perth, Australia, January 1979. Editor: Edwards, C. 1980. 173 figs., 13 tab. VI,487 pages. 830g ⟨3-540-09992-1⟩

Vol. 125: **Nonradial and Nonlinear Stellar Pulsation.** Proceedings of a workshop held at the University of Arizona in Tucson, March 12-16, 1979. Editors: Hill, H. A.; Dziembowski, W. A. 1980. 135 figs., 39 tab. VIII,497 pages. 855g ⟨3-540-09994-8⟩

Vol. 126: **Complex Analysis, Microlocal Calculus and Relativistic Quantum Theory.** Proceedings of the Colloquium Les Houches, Centre de Physique, September 1979. Editor: Iagolnitzer, D. 1980. 22 figs., 4 tab. VIII,502 pages (120 pages in French) 855g ⟨3-540-09996-4⟩

Vol. 127: **Sanchez-Palencia, E.:** Non-Homogeneous Media and Vibration Theory. 1980. 38 figs. IX,398 pages. 700g ⟨3-540-10000-8⟩

Vol. 128: **Neutron Spin Echo.** Laue-Langevin Institut Workshop Grenoble, October 15-16, 1979. Editor: Mezei, F. 1980. VI,253 pages. 455g ⟨3-540-10004-0⟩

Vol. 129: **Geometrical and Topological Methods in Gauge Theories.** Proceedings of the Canadian Mathematical Society Summer Research Institute McGill University, Montréal, September 3-8, 1979. Editors: Harnad, J. P.; Shnider, S. 1980. VIII,155 pages. 295g
⟨3-540-10010-5⟩

Vol. 130: **Mathematical Methods and Applications of Scattering Theory.** Proceedings of a conference held at Catholic University Washington, D. C., May 21-25, 1979. Editors: DeSanto, J. A.; Saenz, A. W.; Zachary, W. W. 1980. 76 figs., 5 tab. XIII,331 pages. 575g
⟨3-540-10023-7⟩

Vol. 131: **Fogedby, H. C.**: Theoretical Aspects of Mainly Low Dimensional Magnetic Systems. 1980. 70 figs. XI,163 pages. 315g
⟨3-540-10238-8⟩

Vol. 132: **Systems Far from Equilibrium.** Sitges Conference on Statistical Mechanics, June 1980, Sitges, Barcelona, Spain. Editor: Garrido, L. 1980. 105 figs., 3 tab. XV,403 pages. 715g
⟨3-540-10251-5⟩

Vol. 133: **Narrow Gap Semiconductors. Physics and Applications.** Proceedings of the International Summer School in Nîmes, France, September 3-15, 1979. Editor: Zawadzki, W. 1980. 304 figs., 22 tab. X,572 pages. 975g
⟨3-540-10261-2⟩

Vol. 134: **Gamma Gamma Collisions.** Proceedings of the International Workshop (Journées d'Etudes Internationales) at Amiens, France, April 8-12, 1980. Editiors: Cochard, G.; Kessler, P. With contributions by numerous experts. 1980. Numerous figs. and tab. XIII, 400 pages. 640g
⟨3-540-10262-0⟩

Vol. 135: **Group Theoretical Methods in Physics.** Proceedings of the IX International Colloquium at Cocoyoc, Mexico, June 23-27, 1980. Editor: Wolf, K. B. 1980. 31 figs., 16 tab. XXVI,629 pages. 1095g
⟨3-540-10271-X⟩

Vol. 136: **The Role of Coherent Structures in Modelling Turbulence and Mixing.** Proceedings of the International Conference Madrid, Spain, June 25-27, 1980. Editor: Jimenez, J. 1981. XIII,393 pages. 680g ⟨3-540-10289-2⟩

Vol. 137: **From Collective States to Quarks in Nuclei.** Proceedings of the Workshop on Nuclear Physics with Real and Virtual Photons, Bologna, Italy, November 25-28, 1980. Editors: Arenhövel, H.; Saruis, A. M. 1981. VII,414 pages. 705g ⟨3-540-10570-0⟩

Vol. 138: **The Many-Body Problem. Jastrow Correlations Versus Brueckner Theory.** Proceedings of the Third Topical School held in Granada (Spain), September 22-27, 1980. Editors: Guardiola, R.; Ros, J. 1981. V,374 pages. 645g
⟨3-540-10577-8⟩

Vol. 139: **Differential Geometric Methods in Mathematical Physics.** Proceedings of the International Conference at the Technical University of Clausthal, Germany, July 1978. Editor: Doebner, H. D. 1981. VII,329 pages. 575g
⟨3-540-10578-6⟩

Vol. 140: **Kramer, P.; Saraceno, M.**: Geometry of the Time-Dependent Variational Principle in Quantum Mechanics. 1981. IV,98 pages. 185g
⟨3-540-10579-4⟩

Vol. 141: **Seventh International Conference on Numerical Methods in Fluid Dynamics.** Proceedings of the Conference, Stanford University, Stanford, California and NASA/Ames (U.S.A.) June 23-27, 1980. Editors: Reynolds, W. C.; MacCormack, R. W. 1981. VIII,485 pages. 830g ⟨3-540-10694-4⟩

Vol. 142: **Recent Progress in Many-Body Theories.** Proceedings of the Second International Conference at Oaxtepec, Mexico, January 12-17, 1981. Editors: Zabolitzky, J. G.; Llano, M. de; Fortes, M.; Clark, J. W. 1981. VIII,479 pages. 830g ⟨3-540-10710-X⟩

Vol. 143: **Present Status and Aims of Quantum Electrodynamics.** Proceedings of the Symposion at Mainz University, May 9-10, 1980. Editors: Gräff, G.; Klempt, E.; Werth, G. 1981. VI,302 pages (26 pages in German) 555g
⟨3-540-10847-5⟩

Vol. 144: **Topics in Nuclear Physics I.** A Comprehensive Review of Recent Developments. Editors: Kuo, T. T. S.; Wong, S. S. M. 1981. XX,567 pages. 965g ⟨3-540-10851-3⟩

Vol. 145: **Topics in Nuclear Physics II.** A Comprehensive Review of Recent Developments. Editors: Kuo, T. T. S.; Wong, S. S. M. 1981. VIII,511 pages. 880g ⟨3-540-10853-X⟩

Vol. 146: **West, B. J.**: On the Simpler Aspect of Nonlinear Fluctuating Deep Water Gravity Waves. (Weak Interaction Theory) 1981. VI,341 pages. 595g ⟨3-540-10852-1⟩

Vol. 147: **Messer, J.**: Temperature Dependent Thomas-Fermi Theory. 1981. IX,131 pages. 260g ⟨3-540-10875-0⟩

Vol. 148: **Advances in Fluid Mechanics.** Proceedings of a Conference Held at Aachen, March 26-28, 1980. Editor: Krause, E. 1981. VII,361 pages. 630g ⟨3-540-11162-X⟩

Vol. 149: **Disordered Systems and Localization.** Proceedings of the conference in Rome, May 1981. Editors: Castellani, C.; Di Castro, C.; Peliti, L. 1981. XII,308 pages. 550g
⟨3-540-11163-8⟩

Vol. 150: **Straumann, N.**: Allgemeine Relativitätstheorie und relativistische Astrophysik. 1981. VII,418 Seiten. 725g ⟨3-540-11182-4⟩

Vol. 151: **Integrable Quantum Field Theories.** Proceedings of the symposium at Tvärminne, Finland, March 23-27, 1981. Editors: Hietarinta, J.; Montonen, C. 1982. V,251 pages. 445g ⟨3-540-11190-5⟩

Vol. 152: **Physics of Narrow Gap Semiconductors.** Proceedings of the 4th International Conference on Physics of Narrow Gap Semiconductors held at Linz, Austria, September 14-17, 1981. Editors: Gornik, E.; Heinrich, H.; Palmetzhofer, L. 1982. XIII,485 pages. 845g ⟨3-540-11191-3⟩

Vol. 153: **Mathematical Problems in Theoretical Physics.** Proceedings of the VIth International Conference in Mathematical Physics, Berlin (West), August 11-20, 1981. Editors: Schrader, R.; Seiler, R.; Uhlenbrock, D. A. 1982. XII,429 pages. 750g ⟨3-540-11192-1⟩

Vol. 154: **Macroscopic Properties of Disordered Media.** Proceedings of a conference held at the Courant Institute, June 1-3, 1981. Editors: Burridge, R.; Childress, S.; Papanicolaou, G. 1982. VII,307 pages. 545g ⟨3-540-11202-2⟩

Vol. 155: **Quantum Optics.** Proceedings of the South African Summer School in Theoretical Physics, held at Cathedral Peak, Natal Drakensberg, South Africa, January 19-30, 1981. Editor: Engelbrecht, C. A. 1982. VIII,329 pages. 575g ⟨3-540-11498-X⟩

Vol. 156: **Resonances in Heavy Ion Reactions.** Proceedings of the symposium held at the Physikzentrum, Bad Honnef, October 12-15, 1981. Editor: Eberhard, K. A. 1982. XII,448 pages. 780g ⟨3-540-11487-4⟩

Vol. 157: **Niyogi, P.**: Integral Equation Method in Transonic Flow. 1982. XI,189 pages. 360g ⟨3-540-11499-8⟩

Vol. 158: **Dynamics of Nuclear Fission and Related Collective Phenomena.** Proceedings of the International Symposium on "Nuclear Fission and Related Collective Phenomena and Properties of Heavy Nuclei" Bad Honnef, Germany, October 26-29, 1981. Editors: David, P.; Mayer-Kuckuk, T.; Woude, A. van der. 1982. X,462 pages. 805g ⟨3-540-11548-X⟩

Vol. 159: **Seiler, E.**: Gauge Theories as a Problem of Constructive Quantum Field Theory and Statistical Mechanics. 1982. V,192 pages. 350g ⟨3-540-11559-5⟩

Vol. 160: **Unified Theories of Elementary Particles.** Critical Assessment and Prospects. Proceedings of the Heisenberg Symposium in München, July 16-21, 1981. Editors: Breitenlohner, P.; Dürr, H. P. 1982. VI,217 pages. 395g ⟨3-540-11560-9⟩

Vol. 161: **Interacting Bosons in Nuclei.** Proceedings of the Fourth Topical School Held in Granada, Spain, September 28 - October 3, 1981. Editors: Dehesa, J. S.; Gomez, J. M. G.; Ros, J. 1982. V,209 pages. 390g ⟨3-540-11572-2⟩

Vol. 162: **Relativistic Action at a Distance: Classical and Quantum Aspects.** Proceedings of the Workshop Held in Barcelona, Spain, June 15-21, 1981. Editor: Llosa, J. 1982. X,263 pages. 475g ⟨3-540-11573-0⟩

Vol. 163: **Darrozes, J. S.; François, C.**: Mécanique des Fluides Incompressibles. 1982. XIX,461 pages. 809g ⟨3-540-11578-1⟩

Vol. 164: **Stability of Thermodynamic Systems.** Proceedings of the meeting held at Bellaterra School of Thermodynamics, Autonomous University of Barcelona, Bellaterra (Barcelona) Spain, September 1981. Editors: Casas-Vázques, J.; Lebon, G. 1982. VII,321 pages. 570g ⟨3-540-11581-1⟩

Vol. 165: **Mukunda, D.; Dam, H. van; Biedenharn, L. C.**: Relativistic Models of Extended Hadrons Obeying a Mass-Spin Trajectory Constraint. Lectures in Mathematical Physics at the University of Texas at Austin. Editors: Böhm, A.; Dollard, J. D. 1982. VI,163 pages. 310g ⟨3-540-11586-2⟩

Vol. 166: **Computer Simulation of Solids.** Editors: Catlow, C. R. A.; Mackrodt, W. C. 1982. XII,320 pages. 570g ⟨3-540-11588-9⟩

Vol. 167: **Fieck, G.**: Symmetry of Polycentric Systems. The Polycentric Tensor Algebra for Molecules. 1982. VI,137 pages. 265g ⟨3-540-11589-7⟩

Vol. 168: **Heavy-Ion Collisions.** Proceedings of the International Summer School La Rábida (Huelva), Spain, June 7-18, 1982. Editors: Madurga, G.; Lozano, M. 1982. VI,429 pages. 740g ⟨3-540-11945-0⟩

Vol. 169: **Sundermeyer, K.:** Constrained Dynamics. With Applications to Yang-Mills Theory, General Relativity, Classical Spin, Dual String Model. 1982. IV,318 pages. 550g
⟨3-540-11947-7⟩

Vol. 170: **Eighth International Conference on Numerical Methods in Fluid Dynamics.** Conference, Rheinisch-Westfälische Technische Hochschule Aachen, Germany, June 28 - July 2, 1982. Editor: Krause, E. 1982. X,569 pages. 965g
⟨3-540-11948-5⟩

Vol. 171: **Time Dependent Hartree-Fock and Beyond.** Proceedings of the International Symposium Held in Bad Honnef, Germany, June 7-11,1982. Editors: Goeke, K.; Reinhard, P.-G. 1982. VIII,426 pages. 735g
⟨3-540-11950-7⟩

Vol. 172: **Ionic Liquids, Molten Salts, and Polyelectrolytes.** Proceedings of the International Conference Berlin (West), June 22-25, 1982. Editors: Bennemann, K.-H.; Brouers, F.; Quitmann, D. 1982. VII,253 pages. 455g
⟨3-540-11952-3⟩

Vol. 173: **Stochastic Processes in Quantum Theory and Statistical Physics.** Proceedings of the International Workshop Marseille, France, June 29 - July 4, 1981. Editors: Albeverio, S.; Combe, P.; Sirugue-Collin, M. 1982. VIII,337 pages. 595g ⟨3-540-11956-6⟩

Vol. 174: **Kadić, A.; Edelen, D. G. B.:** A Gauge Theory of Dislocations and Disclinations. 1983. VII,290 pages. 520g
⟨3-540-11977-9⟩

Vol. 175: **Defect Complexes in Semiconductor Structures.** Proceedings of the International School Held in Mátrafüred, Hungary, September 13-17, 1982. Editors: Giber, J.; Beleznay, F.; Szép, I. C.; László, J. 1983. VI,308 pages. 545g AK ⟨3-540-11986-8⟩

Vol. 176: **Gauge Theory and Gravitation.** Proceedings of the International Symposium on Gauge Theory and Gravitation at Tezukayama University Nara, Japan, August 20-24, 1982. Editors: Kikkawa, K.; Nakanishi, N.; Nariai, H. 1983. X,316 pages. 555g
⟨3-540-11994-9⟩

Vol. 177: **Application of High Magnetic Fields in Semiconductor Physics.** Proceedings of the International Conference Grenoble, France, September 13-17, 1982. Editor: Landwehr, G. 1983. XII,552 pages. 945g ⟨3-540-11996-5⟩

Vol. 178: **Detectors in Heavy-Ion Reactions.** Proceedings of the Symposium Commemorating the 100th anniversary of Hans Geiger's birth, held at the Hahn-Meitner-Institut für Kernforschung Berlin, October 6-8, 1982. Editor: Oertzen, W. von. 1983. VIII,258 pages. 465g
⟨3-540-12001-7⟩

Vol. 179: **Dynamical Systems and Chaos.** Proceedings of the Sitges Conference on Statistical Mechanics, Sitges, Barcelona/Spain, September 5-11, 1982. Editor: Garrido, L. 1983. XIV,298 pages. 545g ⟨3-540-12276-1⟩

Vol. 180: **Group Theoretical Methods in Physics.** Proceedings of the XIth International Colloquium held at Bogazici University, Istanbul, Turkey, August 23-28, 1982. Editor: Serdaroglu, M.; Inönü, E. 1983. XI,569 pages. 975g
⟨3-540-12291-5⟩

Vol. 181: **Gauge Theories of the Eighties.** Proceedings of the Arctic School of Physics 1982, held in Äkäslompolo, Finland, August 1-13, 1982. Editors: Raitio, R.; Lindfors, J. 1983. V,644 pages. 1090g ⟨3-540-12301-6⟩

Vol. 182: **Laser Physics.** Proceedings of the Third New Zealand Symposium on Laser Physics, held at the University of Waikato, Hamilton, New Zealand, January 17-23, 1983. Editors: Harvey, J. D.; Walls, D. F. 1983. V,263 pages. 465g ⟨3-540-12305-9⟩

Vol. 183: **Gunton, J. D.; Droz, M.:** Introduction to the Theory of Metastable and Unstable States. 1983. VI,140 pages. 265g
⟨3-540-12306-7⟩

Vol. 184: **Stochastic Processes.** Formalism and Applications. Proceedings of the Winter School held at the University of Hyderabad, India, December 15-24, 1982. Editors: Agarwal, G. S.; Dattagupta, S. 1983. VI,324 pages. 570g ⟨3-540-12326-1⟩

Vol. 185: **Shirer, H. N.; Wells, R.:** Mathematical Structure of the Singularities at the Transitions Between Steady States in Hydrodynamic Systems. 1983. XI,276 pages. 500g
⟨3-540-12333-4⟩

Vol. 186: **Critical Phenomena.** Proceedings of the Summer School held at the University of Stellenbosch, South Africa, January 18-29, 1982. Editor: Hahne, F. J. W. 1983. VII,353 pages. 620g ⟨3-540-12675-9⟩

Vol. 187: **Density Functional Theory.** Editors: Keller, L.; Gázques, J. L. With contributions by Amador, C.; Das, M. P.; Donnelly, R. A.; Gázquez, J. L.; Harriman, J. E.; Keller, J.; Levy, M.; Perdew, J. P.; Robledo, A.; Varea, C.; Zaremba, E. 1983. V,301 pages. 530g
⟨3-540-12721-6⟩

Vol. 188: **Gauge Symmetries and Fibre Bundles.** Applications to Particle Dynamics. By Balachandran, A. P.; Marmo, G.; Skagerstam, B.-S.; Stern, A. 1983. IV,140 pages. 265g
⟨3-540-12724-0⟩

Vol. 189: **Nonlinear Phenomena.** Proceedings of the CIFMO School and Workshop at Oaxtepec, México, November 29 - December 17, 1982. Editor: Wolf, K. B. 1983. XII,453 pages. 785g
⟨3-540-12730-5⟩

Vol. 190: **Kraus, K.:** States, Effects, and Operations. Fundamental Notions of Quantum Theory. Lectures in Mathematical Physics at the University of Texas at Austin. Editors: Böhm, A.; Dollard, J. D.; Wootters, W. H. 1983. IX,151 pages. 290g ⟨3-540-12732-1⟩

Vol. 191: **Photon Photon Collisions.** Proceedings of the Fifth International Workshop on Photon Photon Collision, held at the Rheinisch-Westfälische Technische Hochschule Aachen, April 13-16, 1983. Editor: Berger, C. 1983. V,417 pages. 725g ⟨3-540-12691-0⟩

Vol. 192: **Heidelberg Colloquium on Spin Glasses.** Proceedings of a Colloquium held at the University of Heidelberg, May 30 - June 3, 1983. Editors: Hemmen, J. L. van; Morgenstern, I. 1983. VII,356 pages. 620g
⟨3-540-12872-7⟩

Vol. 193: **Cool Stars, Stellar Systems, and the Sun.** Proceedings of the Third Cambridge Workshop on Cool Stars, Stellar Systems, and the Sun in Cambridge, Massachusetts, October 5-7, 1983. Editor: Baliunas, S. L.; Hartmann, L. 1984. VII,364 pages. 635g
⟨3-540-12907-3⟩

Vol. 194: **Pascual, P.; Tarrach, R.:** QCD: Renormalization for the Practitioner. 1984. V,277 pages. 490g ⟨3-540-12908-1⟩

Vol. 195: **Trends and Applications of Pure Mathematics to Mechanics.** Invited and Contributed Papers presented at a Symposium at Ecole Polytechnique, Palaiseau, France November 28 - December 2, 1983. Editors: Ciarlet, P. G.; Roseau, M. 1984. V,422 pages (70 pages in French) 725g ⟨3-540-12916-2⟩

Vol. 196: **WOPPLOT 83. Parallel Processing: Logic, Organization, and Technology.** Proceedings of a workshop held at the Federal Armed Forces University, Munich, Neubiberg, Bavaria, Germany, June 27-29, 1983. Editors: Becker, J. D.; Eisele, I. 1984. V,189 pages. 350g ⟨3-540-12917-0⟩

Vol. 197: **Quarks and Nuclear Structure.** 3rd Klaus Erkelenz Symposium Bad Honnef, June 13-16, 1983. Editor: Bleuler, K. 1984. VIII,414 pages. 715g ⟨3-540-12922-7⟩

Vol. 198: **Recent Progress in Many-Body Theories.** Proceedings of the Third International Conference on Recent Progress in Many-Body Theories, Odenthal-Altenberg, Germany August 29 - September 3, 1983. Editors: Kümmel, H.; Ristig, M. L. 1984. IX,422 pages. 735g ⟨3-540-12924-3⟩

Vol. 199: **Recent Developments in Nonequilibrium Thermodynamics.** Meeting at Bellaterra School of Thermodynamics Autonomous University of Barcelona Bellaterra (Barcelona) Spain September 26-30, 1983. Editor: Casas-Vazquez, J.; Jou, D.; Lebon, G. 1984. XIII,485 pages. 845g ⟨3-540-12927-8⟩

Vol. 200: **Zeh, H. D.:** Die Physik der Zeitrichtung. 1984. V,86 Seiten. 180g
⟨3-540-13336-4⟩

Vol. 201: **Group Theoretical Methods in Physics.** Proceedings of the XIIth International Colloquium held at the International Centre for Theoretical Physics, Trieste, Italy September 5-11, 1983. Editors: Denardo, G.; Ghirardi, G.; Weber, T. 1984. XXVII,518 pages. 915g ⟨3-540-13335-6⟩

Vol. 202: **Asymptotic Behavior of Mass and Spacetime Geometry.** Conference at the Oregon State University Corvallis, Oregon, USA October 17-21, 1983. Editor: Flaherty, F. J. 1984. VI,213 pages. 390g ⟨3-540-13351-8⟩

Vol. 203: **Marchioro, C.; Pulvirenti, M.:** Vortex Methods in Two-Dimensional Fluid Dynamics. 1984. III,137 pages. 255g
⟨3-540-13352-6⟩

Vol. 204: **Waseda, Y.:** Novel Application of Anomalous (Resonance) X-ray Scattering for Structural Characterization of Disordered Materials. 1984. VI,183 pages. 335g
⟨3-540-13359-3⟩

Vol. 205: **Solutions of Einstein's Equations: Techniques and Results.** International Seminar Retzbach, Germany, November 14-18, 1983. Editors: Hoenselaers, C.; Dietz, W. 1984. VI,439 pages. 750g ⟨3-540-13366-6⟩

Vol. 206: **Static Critical Phenomena in Inhomogeneous Systems.** Proceedings of the XX Karpacz Winter School of Theoretical Physics, February 20 - March 3, 1984, Karpacz, Poland. Editors: Pekalski, A.; Sznajd, J. 1984. VIII,376 pages. 620g ⟨3-540-13369-0⟩

Vol. 207: **Koch, S. W.:** Dynamics of First-Order Phase Transitions in Equilibrium and Nonequilibrium Systems. 1984. III,148 pages. 285g ⟨3-540-13379-8⟩

Vol. 208: **Supersymmetry and Supergravity /Nonperturbative QCD.** Proceedings of the Winter School Held in Mahabaleshwar, India, January 5-19, 1984. Editors: Roy, P.; Singh, V. 1984. VI,389 pages. 672g ⟨3-540-13390-9⟩

Vol. 209: **Mathematical and Computational Methods in Nuclear Physics.** Sixth Granada Workshop held in Granada, Spain, October 3-8, 1983. Editors: Dehesa, J. S.; Gomez, J. M. G.; Polls, A. 1984. V,276 pages. 490g ⟨3-540-13392-5⟩

Vol. 210: **Cellular Structures in Instabilities.** Proceedings of the Meeting "Structures cellulaires dans les instabilités-périodicité, défauts, turbulence de phase", held at Gif-sur-Yvette (France), June 20-22, 1983. Editors: Wesfreid, J. E.; Zaleski, S. 1984. VI,389 pages. 675g ⟨3-540-13879-X⟩

Vol. 211: **Resonances - Models and Phenomena.** Proceedings of a workshop held at the Centre for Interdisciplinary Research Bielefeld University, Bielefeld, Germany, April 9-14, 1984. Editors: Albeverio, S.; Ferreira, L. S.; Streit, L. 1984. VI,360 pages. 620g ⟨3-540-13880-3⟩

Vol. 212: **Gravitation, Geometry, and Relativistic Physics.** Proceedings of the "Journées Relativistes" held at Aussois, France, May 2-5, 1984. Editor: Laboratoire "Gravitation et Cosmologie Relativistes" Université Pierre et Marie Curie et C. N. R. S. Institut Henri Poincaré, Paris. 1984. VI,336 pages. 585g ⟨3-540-13881-1⟩

Vol. 213: **Forward Electron Ejection in Ion Collisions.** Proceedings of a Symposium at the Physics Institute, University of Aarhus, Denmark, June 29-30, 1984. Editors: Groeneveld, K. O.; Meckbach, W.; Sellin, I. A. 1984. VII,165 pages. 310g ⟨3-540-13887-0⟩

Vol. 214: **Moraal, H.:** Classical, Discrete Spin Models: Symmetry, Duality, and Renormalization. 1984. VII,251 pages. 455g ⟨3-540-13896-X⟩

Vol. 215: **Computing in Accelerator Design and Operation.** Proceedings of the Europhysics Conference held at the Hahn-Meitner-Institut für Kernforschung Berlin GmbH, September 20-23, 1983. Editors: Busse, W.; Zelazny, R. 1984. XII,574 pages. 985g ⟨3-540-13909-5⟩

Vol. 216: **Applications of Field Theory to Statistical Mechanics.** Proceedings of the Sitges Conference on Statistical Mechanics Sitges, Barcelona /Spain, June 10-15, 1984. Editor: Garrido, L. 1985. VIII,352 pages. 620g ⟨3-540-13911-7⟩

Vol. 217: **Charge Density Waves in Solids.** Proceedings of the International Conference held in Budapest, Hungary, September 3-7, 1984. Editors: Hutiray, G.; Sólyom, J. 1985. XIV,541 pages. 935g ⟨3-540-13913-3⟩

Vol. 218: **Ninth International Conference on Numerical Methods in Fluid Dynamics.** Editors: Soubbaramayer; Boujot, J. P. 1985. X,612 pages. 1035g ⟨3-540-13917-6⟩

Vol. 219: **Fusion Reactions Below the Coulomb Barrier.** Proceedings of an International Conference held at the Massachusetts Institute of Technology, Cambridge, MA, June 13-15, 1984. Editor: Steadman, S. G. 1985. VII,351 pages. 620g ⟨3-540-13918-4⟩

Vol. 220: **Dittrich, W.; Reuter, M.:** Effective Lagrangians in Quantum Electrodynamics. 1985. V,244 pages. 440g ⟨3-540-15182-6⟩

Vol. 221: **Quark Matter '84.** Proceedings of the Fourth International Conference on Ultra-Relativistic Nucleus-Nucleus Collisions Helsinki, Finland, June 17-21, 1984. Editor: Kajantie, K. VI,305 pages. 1985. 545g ⟨3-540-15183-4⟩

Vol. 222: **Garcia, A.; Kielanowski, P.:** The Beta Decay of Hyperons. Lectures in Mathematics and Physics at the University of Texas at Austin. Foreword and addendum by Bohm, A. 1985. VIII,173 pages. 325g⟨3-540-15184-2⟩

Forthcoming Volumes

Vol. 223: **Saller, H.:** Vereinheitlichte Feldtheorie der Elementarteilchen. Eine Einführung. 1985. 157 pages ⟨3-540-15188-5⟩

Vol. 224: **Supernovae as Distance Indicators.** Proceedings of a Workshop Held at the Harvard-Smithsonian Center for Astrophysics, September 27-28, 1984. Editor: N. Bartel. 1985. 226 pages ⟨3-540-15206-7⟩

Vol. 225: **Müller, B.:** The Physics of the Quark-Gluon Plasma. 1985. 142 pages ⟨3-540-15211-6⟩

Vol. 226: **Non-Linear Equations in Classical and Quantum Field Theory.** Proceedings of a Seminar Series Held at DAPHE, Observatoire de Meudon, and LPTHE, Université Pierre et Marie Curie, Paris, between October 1983 and October 1984. Editor: N. Sanchez. 1985. 400 pages ⟨3-540-15213-X⟩

Vol. 227: **Eckmann, J.-P., Wittwer, P.:** Computer Methods and Borel Summability Applied to Feigenbaum's Equation. 1985. 297 pages ⟨3-540-15215-6⟩

Vol. 228: **Thermodynamics and Constitutive Equations.** Lectures Given at the 2nd 1982 Session of the Centro International Matematico Estivo (C.I.M.E.) Held at Noto, Italy, June 23 – July 2, 1982. Editor: G. Grioli. 1985. 257 pages ⟨3-540-15228-8⟩

Vol. 229: **Fundamentals of Laser Interactions.** Proceedings of a Seminar Held at Obergurgl, Austria, February 24 – March 2, 1985. Editor: F. Ehlotzky. 1985. 314 pages ⟨3-540-15640-2⟩

Vol. 230: **Macroscopic Modelling of Turbulent Flows.** Proceedings of a Workshop Held at INRIA, Sophia-Antipolis, France, December 10-14, 1984. Editors: U. Frisch, J.B. Keller, G.C. Papanicolaou, O. Pironneau. 1985. 370 pages ⟨3-540-15644-5⟩

Vol. 231: **Hadrons and Heavy Ions.** Proceedings of the Summer School Held at the University of Cape Town, January 16-27, 1984. Editor: W.D. Heiss. 1985. 465 pages ⟨3-540-15653-4⟩

Vol. 232: **New Aspects of Galaxy Photometry.** Proceedings of the Specialized Meeting of the Eighth IAU European Regional Astronomy Meeting, Toulouse, France, September 17-21, 1984. Editor: J.-L. Nieto. 1985. 350 pages ⟨3-540-15657-7⟩

Vol. 233: **High Resolution in Solar Physics.** Proceedings of a Specialized Session of the Eighth IAU European Regional Astronomy Meeting, Toulouse, France, September 17-21, 1984. Editor: R. Muller. 1985. 320 pages ⟨3-540-15678-X⟩

Vol. 234: **Electron and Photon Interactions at Intermediate Energies.** Proceedings of the 1984 Workshop, Held at Bad Honnef, Germany, October 29-31, 1984. Editors: D. Menze, W. Pfeil, W.J. Schwille. 1985. 481 pages ⟨3-540-15787-9⟩

Vol. 235: **Flow of Real Fluids.** Editors: G.E.A. Meier, F. Obermeier. 1985. 348 pages ⟨3-540-15989-4⟩

Vol. 236: **Advanced Methods in the Evaluation of Nuclear Scattering Data.** Proceedings of the International Workshop Held at the Hahn-Meitner-Institut für Kernforschung Berlin, June 18-20, 1985. Editors: H.J. Krappe, R. Lipperheide. 1985. 364 pages ⟨3-540-15990-8⟩

Vol. 237: **Nearby Molecular Clouds.** Proceedings of a Colloquium of the Eighth IAU European Regional Astronomy Meeting, Toulouse, France, September 17-21, 1984. Editor: G. Serra. 1985. 238 pages ⟨3-540-15991-6⟩

Vol. 238: **The Free-Lagrange Method.** Proceedings of the First International Conference on Free Lagrange Methods Held at Hilton Head Island, South Carolina, March 1985. Editors: M.J. Fritts, W.P. Crowley, H. Trease. 315 pages ⟨3-540-15992-4⟩

2. Author Index

44

II. The Publication Process

To reach the goal of rapid publication at a low price it was necessary to simplify the production especially to avoid the many intermediate steps otherwise common in book production like copy editing, type setting, proofreading of galleys, etc. So, the technique of photographic reproduction from a camera-ready manuscript was chosen.

This process shifts the main responsibility for the technical quality considerably from the publisher to the author. Therefore we would like to urge the authors and also the editors of proceedings to observe very carefully the instructions to be found below. Also, it might be useful to look into some of the volumes already published or, especially if some non-typical text is planned, to write to the Physics Editorial of Springer-Verlag directly. This not only avoids mistakes and time-consuming correspondence during the production period, but it also will help the author to make full use of one of the greatest advantages of this technique of production, namely, that the layout is almost entirely in his hands and no printer will have a chance to blur his formulae or to improve his style.

Editors of multi-authored volumes should also study the rules for the publication of conference proceedings to be found below and apply them by analogy to their publications.

Note: Careful editing helps to avoid delays and results in a publication time of six to eight weeks after submission of the complete manuscript for production.

1. How to Organize the Publication of Conference Proceedings

Proposals should be submitted to the physics editor at Springer-Verlag who will then forward them to the appropriate series editor. Upon request, Springer-Verlag will furnish the name of the series editor, so that he may be contacted directly. In this case, copies of all letters should be sent to Springer-Verlag.

When applying for publication in the series **LECTURE NOTES IN PHYSICS** the volume's editor(s) should submit enough material to permit the series editors and their referees to make a fairly accurate evaluation (a complete list of speakers and titles of papers to be presented, abstracts, a rough estimate of the number of pages, etc.). If, based on this information, the proceedings are accepted, the volume's editor(s) whose name(s) will appear on the title page will have to take over: He (they) will select the papers suitable for publication and will have them refereed when appropriate. The series editors will normally not interfere with the detailed editing except in fairly obvious cases or on technical matters.

The proceedings should be limited to only a few areas of research and those should be closely related to each other. The contributions should be of a high standard, of current interest and should avoid lengthy redraftings of papers already published elsewhere. As a whole, the proceedings should aim for a balanced presentation of the theme of the conference including a description of the techniques used and enough motivation for the reader to arrive at the results himself. A table of contents has to be included and a descriptive introduction would be helpful. As a rule, discussions will not be accepted; a listing of papers presented at the meeting but not included in the proceedings would be appreciated.

To avoid the publication of outdated papers the manuscript should be forwarded to the publisher no later than four months after the meeting. In cases of extreme delay the series editors should check once more the timeliness of the papers before the manuscript is sent to the publisher. Therefore, the volume's editor(s) should establish strict deadlines or, in case of delayed publication, should encourage the authors to update their contributions if appropriate. The most efficient way, by far, to speed up the publication is to ask the speakers to bring their articles to the conference and to supply them with secretarial help for last-minute changes during the meeting.

The volume's editor(s) will receive 50 free copies (**not** for resale). In case he (they) wish(es) to order additional copies they should contact the publisher for information. Unfortunately, the standard production process does not allow the publisher to supply the authors with reprints of their contributions.

Upon request, Springer-Verlag will supply the volume's editor(s) with special stationery and typing instructions for distribution among the speakers. Contributions badly typed with hardly readable formulae or figures and photographs not suitable for reproduction should be rejected, otherwise the publisher must send them back for retyping or ask the author(s) for better figures or photographs, respectively, thereby delaying the production process considerably.

Before sending the original manuscript to Springer-Verlag, please take note of the following points:

a) Pages should be numbered in blue pencil only. Pagination of the preface, table of contents, and other front-matter will be completed by Springer-Verlag. The first page of the first contribution is page 1. If you intend to divide the volume into different groups, the title page for each group should be a right-hand one (odd-numbered page). Thus, the text of the first contribution starts on page 3 (page 2 remains vacant as the back page of the first title page); the back pages of the following title pages need to be taken into account accordingly.

Please make sure that the manuscript contains no other pagination. If necessary, page numbers inserted by authors of contributions should be removed with white correction fluid.

b) The table of contents must correspond **exactly** to the titles and the names of contributors as they appear in the contributions. If necessary, minor corrections can be made by Springer-Verlag, where various IBM golf-ball type heads are available for this purpose, e.g., Letter Gothic, Courier 10, Courier 12, Delegate. However, Springer-Verlag is forced to return material typed with other kinds of type or typewriters to the editor(s) for any necessary corrections.

c) All prefaces will be checked by an English copy editor for language and style. Point b) will apply in cases of any corrections.

Final acceptance is generally expressed by the series editor in charge, in agreement with Springer-Verlag, after receiving the complete manuscript. Therefore, a copy of the manuscript should be sent to the series editor as soon as possible so that any necessary changes can be discussed with him. As a general rule, the series editor confirms his preliminary acceptance of the manuscript on the basis of the abstracts or the table of contents if

a) the final contents of the articles correspond to the originally discussed concept;

b) the quality of the contributions meets the requirements of the series;

c) the final version of the contributions does not greatly exceed the numbers of pages originally agreed on.

Grave linguistic or technical shortcomings may lead to the rejection of contributions.

For authors of monographs everything stated above regarding submission of manuscripts, free copies and technical matters applies accordingly. For further information, please contact Springer-Verlag's Physics Editorial.

2. Essentials for the Preparation of Camera-Ready Manuscripts

Fifteen years of experience have shown that best results can be achieved and unnecessary delays can be avoided, if the instructions given below are followed strictly.

Typing area. Springer-Verlag will supply, on request, special paper with the typing area outlined.
The correct typing area is 18 x 26 1/2 cm (7 x 10 1/2 inches).

Make sure the TYPING AREA IS COMPLETELY FILLED. Set the margins so that they precisely match the outline, and type right from the top to the bottom line. Lines of text should not end more than three spaces inside or outside the right margin (see example).

Type on one side of the paper only.

Type. Use an electric typewriter if at all possible. CLEAN THE TYPE before use and always use a BLACK ribbon (a carbon ribbon is best).

Choose a type size large enough to stand reduction to 75%.

Spacing, Headings. Use ONE-AND-A-HALF line spacing in the text.

Start each chapter or paper on a NEW PAGE and leave sufficient space for the title to stand out clearly. However, do NOT use a new page for the beginning of subdivisions of chapters. Leave THREE LINES blank above and TWO below headings of such subdivisions.

Make sure headings of equal importance are in the same form.

Footnotes. These are frowned upon. In exceptional cases, place them at the foot of the page, separated from the text by a line 4 cm long, and type them in SINGLE LINE spacing to finish exactly on the outline.

Symbols. Anything which cannot be typed may be entered by hand in BLACK AND ONLY BLACK ink. (A fine-tipped rapidograph is suitable for this purpose; a good black ball-point will do, but a pencil will not.) Do not draw straight lines by hand without a ruler (not even in fractions). However, always type as much as possible. If you are using an IBM typewriter, fit the "SYMBOL" golf-ball.

Emphasis. Words to be emphasized can either be UNDERLINED or typed in ITALICS.

Literature references. These should be placed at the end of each paper or chapter, or at the end of the work, as desired. Type them in single line spacing, and start each reference on a new line. Follow "Zentralblatt Mathematik" for titles of mathematical journals and "Bibliographic Guide for Editors and Authors (BGEA)" for chemical, physical and biological journals.

Figures. Mark spaces in the text to fit any FIGURES to be inserted, and add two lines of spacing above and below them.
Type LEGENDS TO FIGURES in single line spacing, and place them on the page below the figures. Leave three blank lines before continuing the running text.

Use for the FIGURES either ORIGINAL LINE DRAWINGS or POSITIVE COPIES. Check that lines and points on copies are UNIFORMLY BLACK. Paste them into the spaces left for this purpose. (When positive copies are used, we would appreciate your sending us the originals, if available, enclosed in a separate envelope.) Letters or numerals ON figures should be about 3 mm high, but in any case no smaller than the characters on the typewriter.

Where photographs (half-tones) are to be reproduced, leave a space as described. ENCLOSE the photographs with the text. They should be glossy prints with evenly black lines and points in the size required or better still enlarged. MARK the space in the text and the back of the photograph CLEARLY so that there cannot possibly be any doubt about where it should be placed or which way up. Color photographs are reproduced in black and white only. If reproduction in color is desirable, please contact the publisher.

IMPORTANT

Pagination. Number pages in the upper right-hand corner or on the back in BLUE CRAYON ONLY. The final pagination will be done by Springer-Verlag.

It is much safer to number pages after the text has been typed and corrected. Page 1 should be THE FIRST PAGE OF THE ACTUAL TEXT. Preface, contents page, abstract, acknowledgements, brief introduction, etc., should NOT be numbered.

Corrections. When corrections have to be made, cut the new text to fit, and PASTE it over the old. White correction fluid may also be used.

NEVER make corrections or insertions in the text by hand.

If the typewriter has to be marked for any reason, e.g. for TEMPORARY page numbers or to mark corrections for the typist, this can be done VERY FAINTLY with BLUE or GREEN CRAYON but NO OTHER COLOR.

Table of contents. It is advisable to type the table of contents last, copying the titles from the text.

Finally a word about PACKING. To avoid damage in the mail, cover the manuscript with cardboard.

Example 1:

MEASUREMENT AND IMPROVEMENT OF MEMORY ALLOCATION

IN A PROCESS COMPUTER

H.A. Spang
General Electric Research
and Development Center
Schenectady, NY 12301/USA

A good example of a heading to a conference report. It contains all necessary information.

Example 2:

V and W are white Gaussian random vectors sequences, with :

$$E[x_o] = \hat{x}_o \quad , \quad E[x_o x_o^T] = P_o \qquad I.1$$

$$E[w] = 0 \quad , \quad E[w_k w_j^T] = Q_k \delta_{kj} \qquad I.2$$

where:
$$E[v] = 0 \quad , \quad E[v_k v_j^T] = R_k \delta_{kj} \qquad I.3$$

$$\delta_{kj} = \begin{cases} 0 & , \ k \neq j \\ 1 & , \ k = j \end{cases}$$

is the Kronacker delta and P_o, Q_k and R_k are positive definite

An example of a formula which has to be rejected: The author ignores correct spacing, makes no clear distinction between subscripts and the letters on line, each line bends downwards, the letters are out of focus, etc.

Example 3:

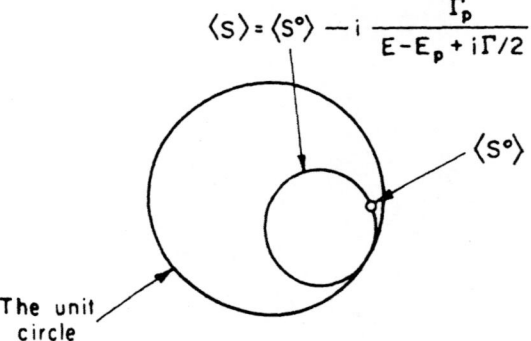

$$\langle S \rangle = \langle S° \rangle - i \, \frac{\Gamma_p}{E - E_p + i\Gamma/2}$$

$\langle S° \rangle$

The unit circle

This is an excellent sample for the size of the letters on enlarged figures. The letters have to be enlarged accordingly.